W9-AFS-763

Take Off with

PUZZLES

About This Book

The activities, puzzles, and games in this book have been designed for an adult and child to enjoy together. Take time to find out about the many opportunities they provide for learning about math, from counting and adding to measuring and sorting.

Each page deals with a topic that children will be introduced to in the early years at school. The book revisits and expands on themes that are covered in other titles in the series, emphasizing the element of play that is so important at this stage of learning.

The "Take Off" loops will give children a chance to develop the skills they have been practicing on that page and to include their friends. Children learn most effectively by joining in, talking, asking questions, and solving problems, so encourage them to talk about what they are doing and to find ways of solving the problems by themselves.

Use the book as a starting point. Look for other occasions to learn about math puzzles; for example, create party games, count and sort the things you see on a special day out, or simply discuss the shapes and positions of everything in the kitchen. Make sure that it is not only easy to take off with puzzles but also fun!

Published by Raintree Steck-Vaughn Publishers, an imprint of Steck-Vaughn Company

Library of Congress Cataloging-in-Publication Data

Hewitt, Sally.
Puzzles / Sally Hewitt.
p. cm. — (Take off with)
ISBN 0-8172-4115-9
1. Mathematical recreations — Juvenile literature. [1. Mathematical recreations.]
I. Title. II. Series: Hewitt, Sally. Take off with.
QA95.H49 1996
648'.68—dc20 95-24259
CIP
AC

Printed in Hong Kong
Bound in the United States
1 2 3 4 5 6 7 8 9 0 LB 99 98 97 96 95

Sally Hewitt

Austin, Texas

Acknowledgments

Editorial: Rachel Cooke, Kathy DeVico
Design: Ann Samuel, Joyce Spicer
Production: Jenny Mulvanny, Scott Melcer
Photography: Michael Stannard
Consultant: Peter Patilla, formerly Senior Lecturer in mathematics education, Sheffield Hallam University
Artwork: Clinton Banbury Associates and Deborah Crow
Model making: Deborah Crow

The author and publishers would like to thank the following companies for their help with the objects photographed for this book:

Early Learning Centre, pages 8, 9 and 12; John Lewis Partnership, pages 8, 9, 12, 13, 14, 15, 16, 22, 23 and 26; NES Arnold Limited, pages 8, 23 and 27; Stockingfillas, pages 8 and 9; Tridias, 6 Bennett Street, Bath BA1 2QP, 01225 314730, pages 8, 9, 12, 13, 14, 17, 22 and 23.

For permission to reproduce copyright material, the author and publishers gratefully acknowledge the following:

Page 29: (top left) Andrea Pistolesi/The Image Bank, (top right) Joe Szkodzinski/The Image Bank.

Contents

Count and Sort

Look at the big collection of objects on these two pages. You can count and sort them in many different ways.

Count all of the legs.
How many ears can you see?
Count the things that you might find in a bathroom.

Find everything that can fly.
Find everything that can swim.
Find two hair ties that are exactly the same.

Look for everything that is colored red.
Find all of the round shapes.
What can you eat with?

Give us a clue

Play this game with a friend.

Choose one of the objects on this page. Don't say which one, but give your friend a hint. For example, "I see something with four legs."

Keep adding clues until your friend guesses the right object.

How Many Fish?

Use the pictures to help you answer the problems and finish the rhymes!

1 + 2 = ?

One little fish calls, "Follow me!"
Two come along. Now there are ?

3 + 3 = ?

Three little fish, getting up to tricks.
Three join the play. Now there are ?

6 + 3 = ?

Six little fish are swimming in a line.
Here come three more. Now there are ?

9 – 4 = ?

Nine little fish are going for a dive.
Four lose their way. Now there are ?

5 – 1 = ?

Five little fish are heading for the shore.
One meets a friend. Now there are ?

4 – 3 = ?

Four little fish are tired from all the fun.
Three swim off home. This leaves only ?

Adding Wheels

Count the wheels on all of these toys.
Then play this adding game.

The bicycle, the airplane, and the wheelbarrow have six wheels altogether. Point to two toys that have four wheels altogether. Now point to two different toys that have four wheels altogether.

Find three toys that have five wheels altogether. Which four toys have nine wheels altogether?

The pink car has four wheels.
Which toys will go with it to make
seven wheels altogether?

Find a group of toys that have
ten wheels altogether.
Find a different group of toys
that have ten wheels altogether.

Can you make up some more questions?

Remember, Remember

Take a good look at everything on this page.
Now cover up the picture.
How many things can you remember?

A memory game

Look at a shelf at home with a friend.
Then turn away from the shelf.

Who can remember the most things on the shelf?

Can you remember where everything is?
Study the picture, cover it up, and then
answer these questions.
If you get stuck, look at the picture
again for a minute, and then
start the questions again.

- What is on the table?
- Are the tennis rackets under the table?
- Where is the bat, next to the table or on top of it?
- What is in front of the red bag, the sneakers or the bowling pins?
- Is there anything inside of the red bag?
- How many bowling pins are behind the balls?
- What is on top of the flippers?
- What color hat is on the bottom peg of the hatstand?

Find the Pattern

Can you match each of these rubbings to the object that it has been taken from?

Make a rubbing

Find something with an interesting pattern that you can feel.

Place a piece of blank paper over it, and rub the paper firmly with a crayon. Watch as a rubbing pattern appears on the paper.

Where should you put these petals to repeat the pattern around the flower once more? Will you need all of them? What color petal is left over?

Look at the ways these groups of toys have been arranged. Which two groups are the same?

Make a new pattern, starting with the red dinosaur. Where would you put the blue and yellow ones?

Can you find any more ways to arrange the toys?

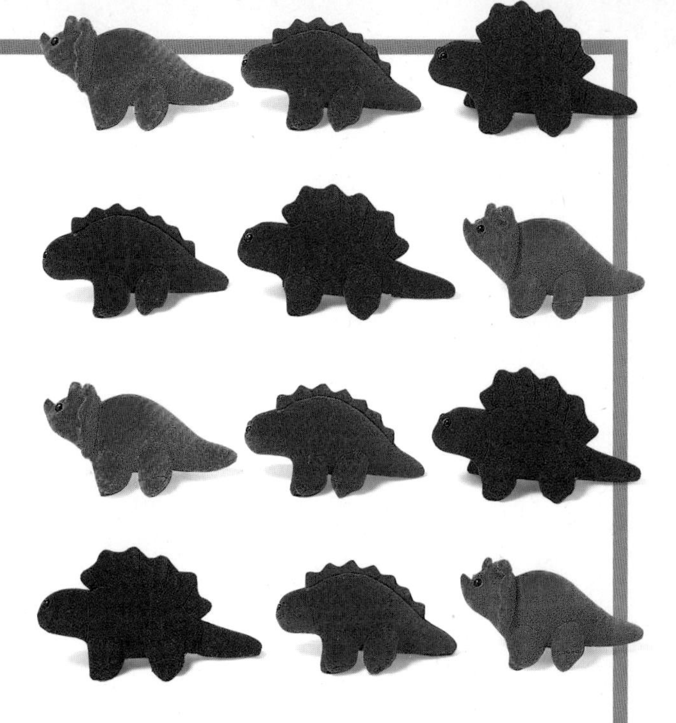

How many strawberries should there be in the missing group?

There should be six. Can you say why?

How many pebbles should there be in the missing group?

?

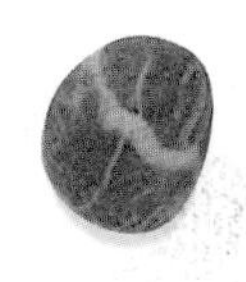

There should be three. Can you say why?

What Happens Next?

Look at these pictures. You will find all the things you need to build a castle, make a party hat, and dress a doll. Say how you would use all of these things to complete each task.

How would you build this castle? In what order would you use each of the blocks that you need?

Describe how you would make this party hat. What would you do first? What would you do last?

In what order would you put Katie's clothes on?
Would you start with the hat?

Can you think of different orders in which you might do each of the tasks on these two pages?

Cartoon action

Cut four small rectangles of paper.
Think of a short story.
Draw one part of the story on each rectangle of paper.

Mix them up, and see if a friend can put them back in the right order.
You could cut up a comic strip and do the same thing.

Space Visitors

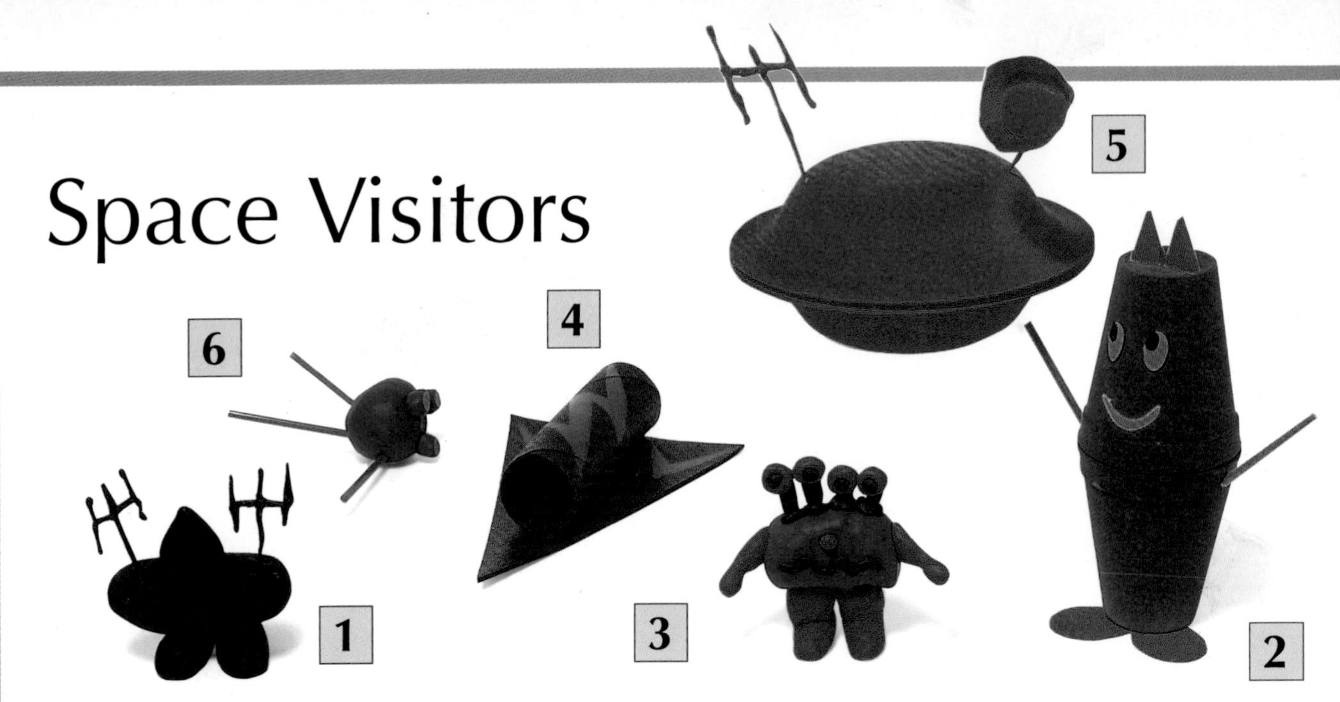

Chris, Pearl, Jane, Jenna, Karim, and Tim have made these space models.

They could choose to make either an alien or a spacecraft.

The models could be made from modeling clay or from old boxes.

They could use red, blue, or yellow paint.

Use the diagrams to figure out who made each model.

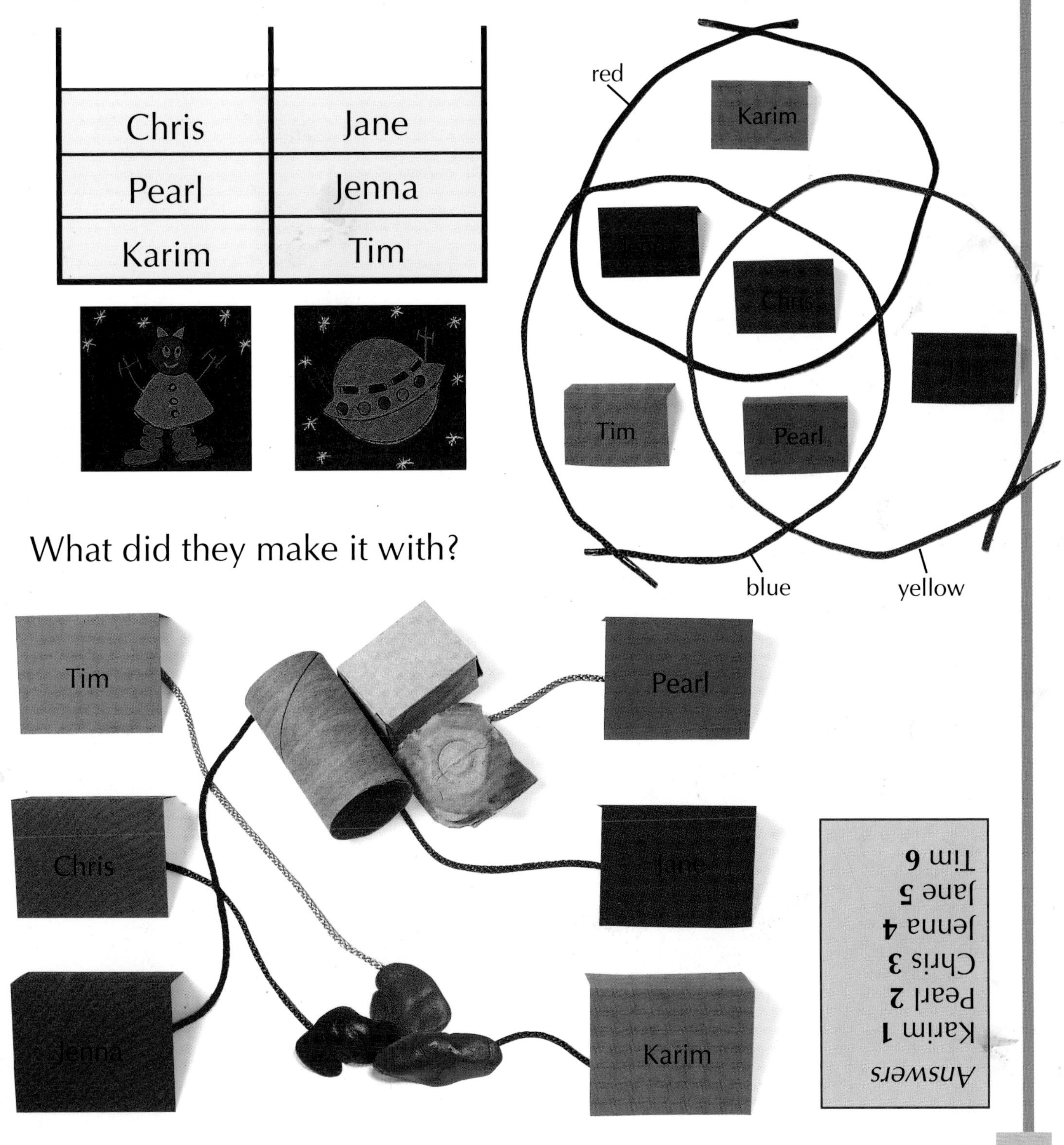

How Has It Moved?

All of these things have either been flipped upside down or turned sideways.

Which ones have been flipped upside down, and which ones have been turned sideways?

One has been turned sideways but still looks the same. Which one is it?

Look at the row of letters.

How do they look different reflected in the mirror?

The pictures on the right of the red line are reflections of the pictures on the left. Or are they? Can you find the mistakes?

You can put a mirror on the red line to find out what you should really see.

Reflected patterns

Fold some paper in half, and then open it up.
Paint a simple pattern on one side of the fold.
Fold the paper in half again, and press it down.
Open the paper up to see your reflected pattern.

Treasure Island

Play these games on Treasure Island.
With each move, say if you are turning left or right, straight or backward.

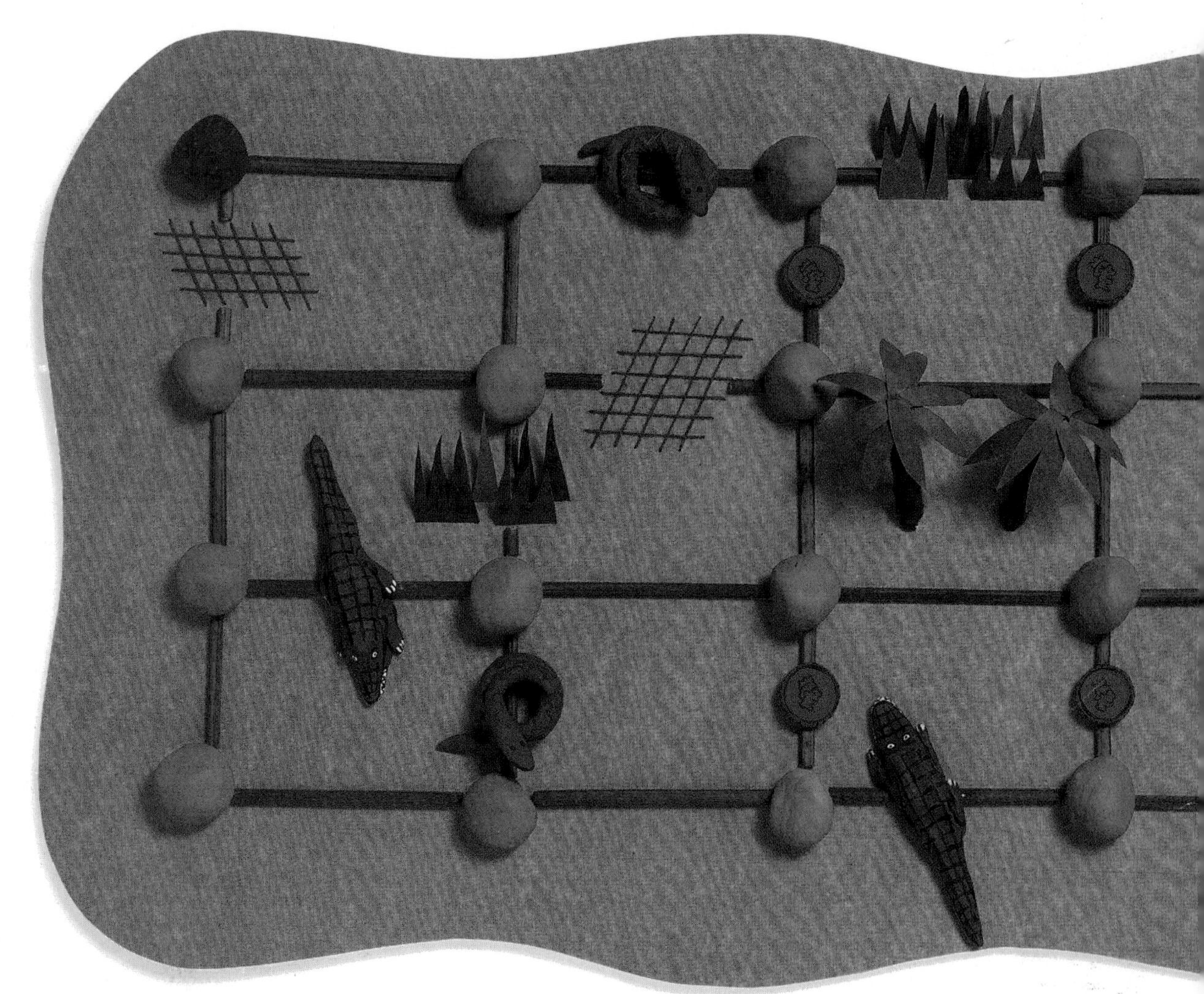

Game 1
Cross the island from one red stone to the other. You can only move upward, downward, or sideways. You cannot go over a snake, a swamp, a crocodile, or a trap. But you can go back to a stone that you have already visited.

Game 2

Play the game again. But this time, collect all four coins as you go, by crossing over them. You can only pick each coin up once, even if you cross over it twice.

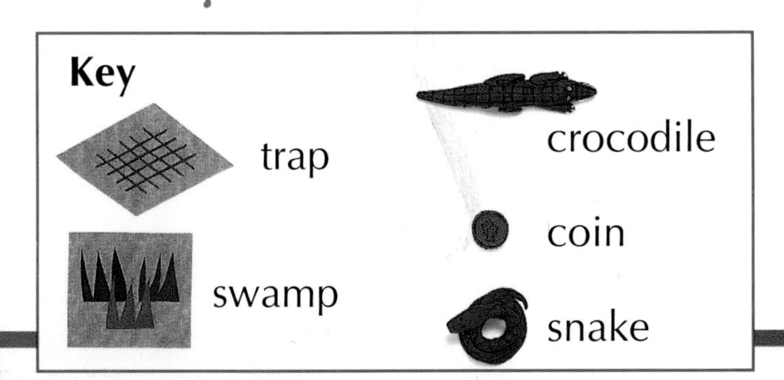

Island race

Play this version of the Treasure Island game with a friend.
It is a race across the island to get to each other's stone.

You need two game pieces and some dice.

- Each start on a red stone.
- Take turns throwing the dice.
- The number you throw is the number of places you can move across the island.
- If you throw a six, you miss a turn!
- If you throw a two, you can leap over a snake, a swamp, a crocodile, or a trap (remember you can't the rest of the time).
- You can't both stand on the same stone at once, even in the middle of a turn, so plan your route carefully.

The winner is the first to pick up all four coins and reach the other side of the island.

Guessing Games

Guess which snail has crawled the farthest.

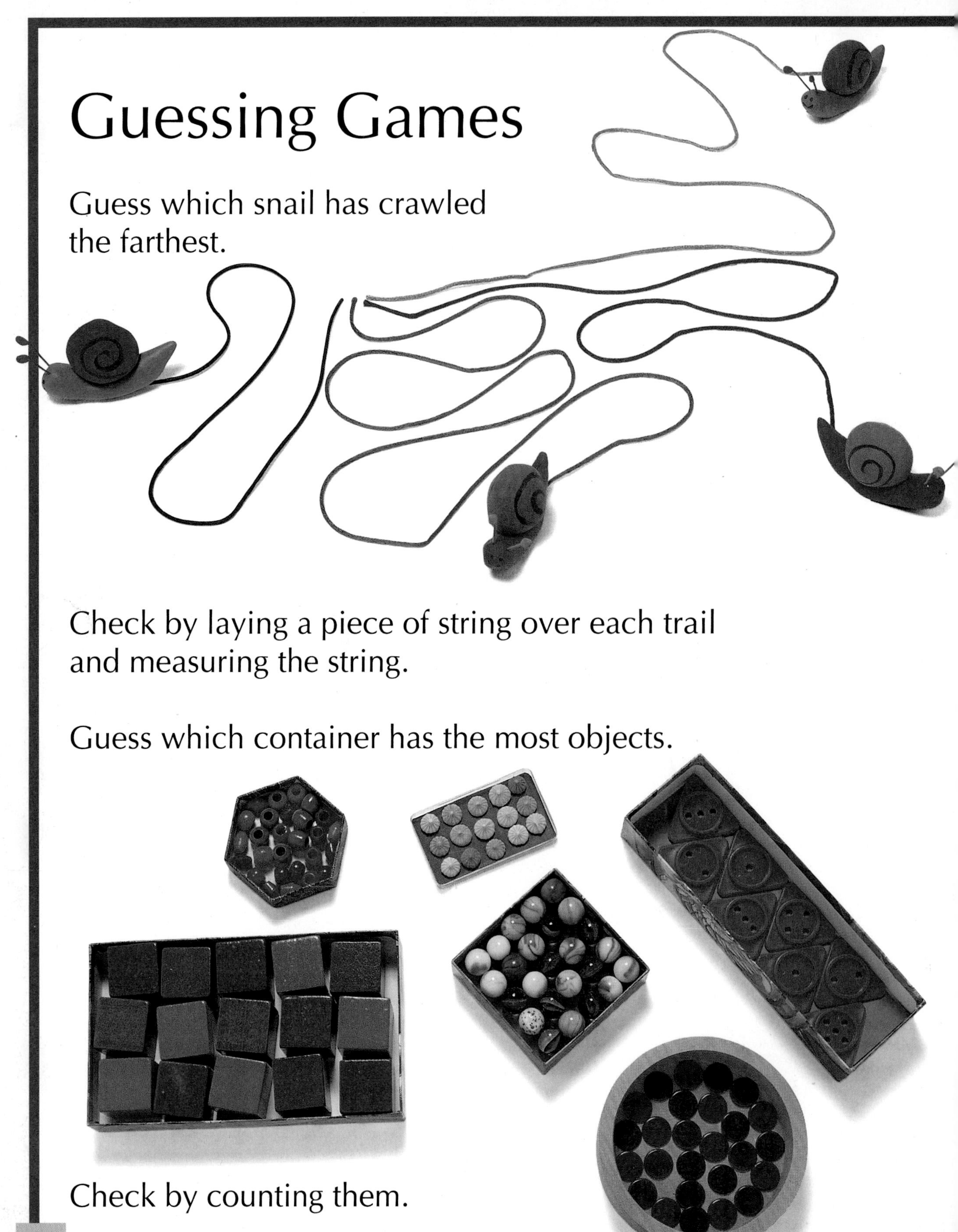

Check by laying a piece of string over each trail and measuring the string.

Guess which container has the most objects.

Check by counting them.

Each one of these patterns has been made by cutting one of the flat shapes beneath them.

Guess which shape has been used to make which pattern.

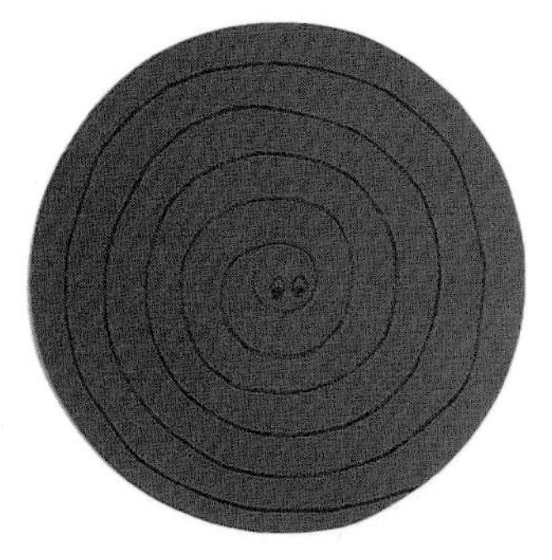

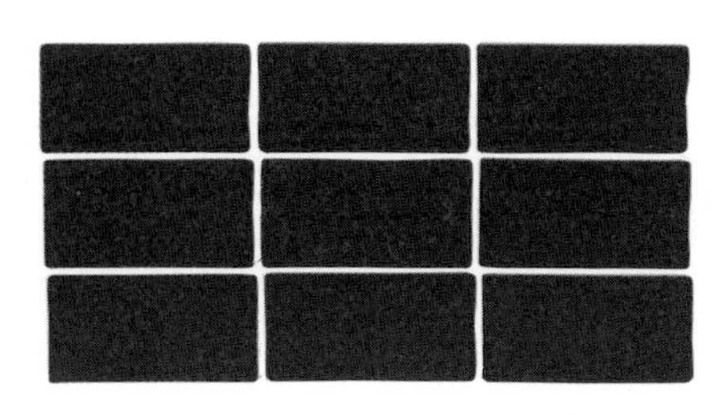

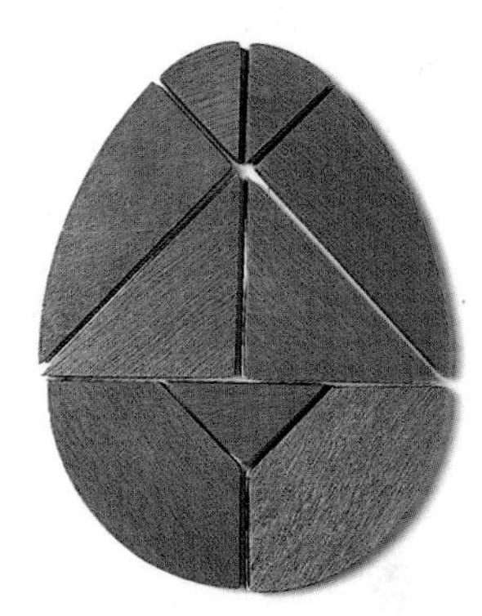

Check by cutting shapes and making the patterns.

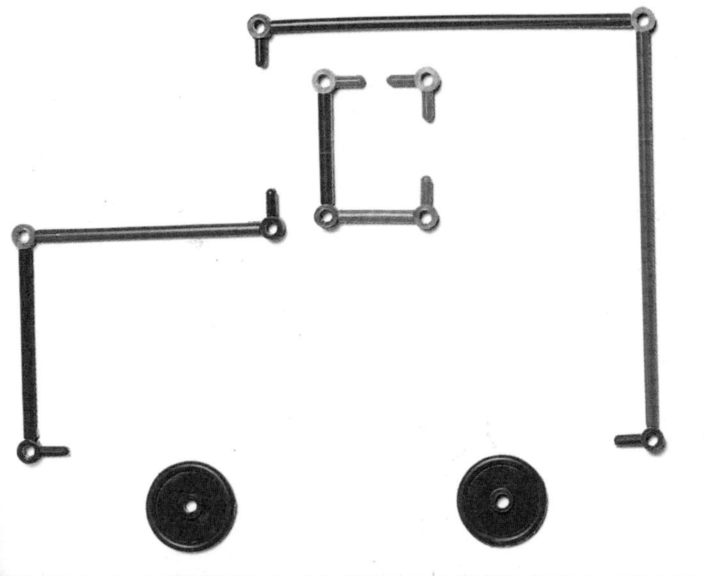

Guess where these missing straws fit in the truck. Check by measuring the pieces and then measuring the gaps.

Shapes Within Shapes

Sometimes two or more shapes put together make another shape.

What do these four triangles make?

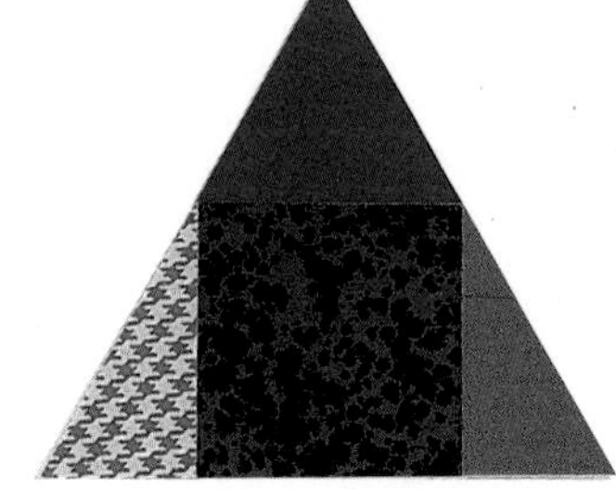

What shapes make up this triangle?

Can you make a diamond with these triangles?

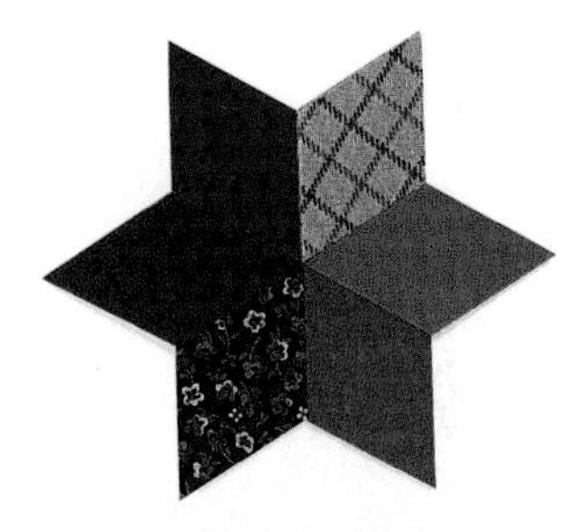

How many diamonds make this star?
How many triangles do you need to make the same-shaped star?

Look for these shapes within shapes in this pattern:

- a rectangle made by two triangles
- a five-sided shape made by two other shapes
- a six-sided shape made by four other shapes
- a rectangle made by 20 other shapes

What shapes can you see in these two pictures?

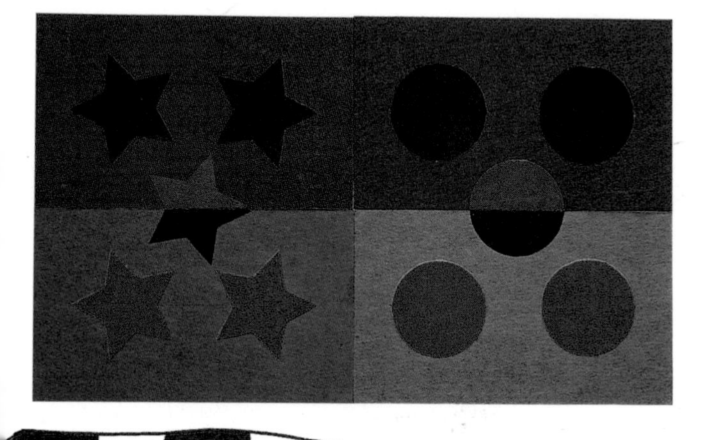

This big rectangle is made from four small rectangles. But can you see four more rectangles in the shape, each one made from two of the small rectangles?

A square challenge

Make nine cardboard squares. They should be the same size but different colors. Lay them down to make a big square, like this one.

Now challenge your friends to say how many squares they can see. They should say 14! These squares made from four squares give you a hint why.

Pattern Puzzles

Look at the patterns.
Which shape comes next?

1.

2.
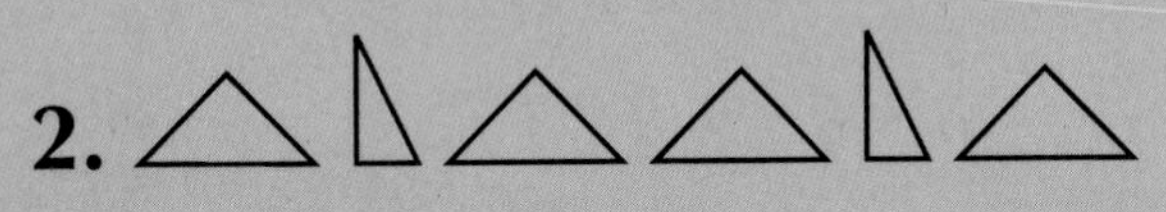

3. □ ◺ ◇ □ ◺ ◇ □

Look at each group of shapes.
How many shapes should there be in the missing group?

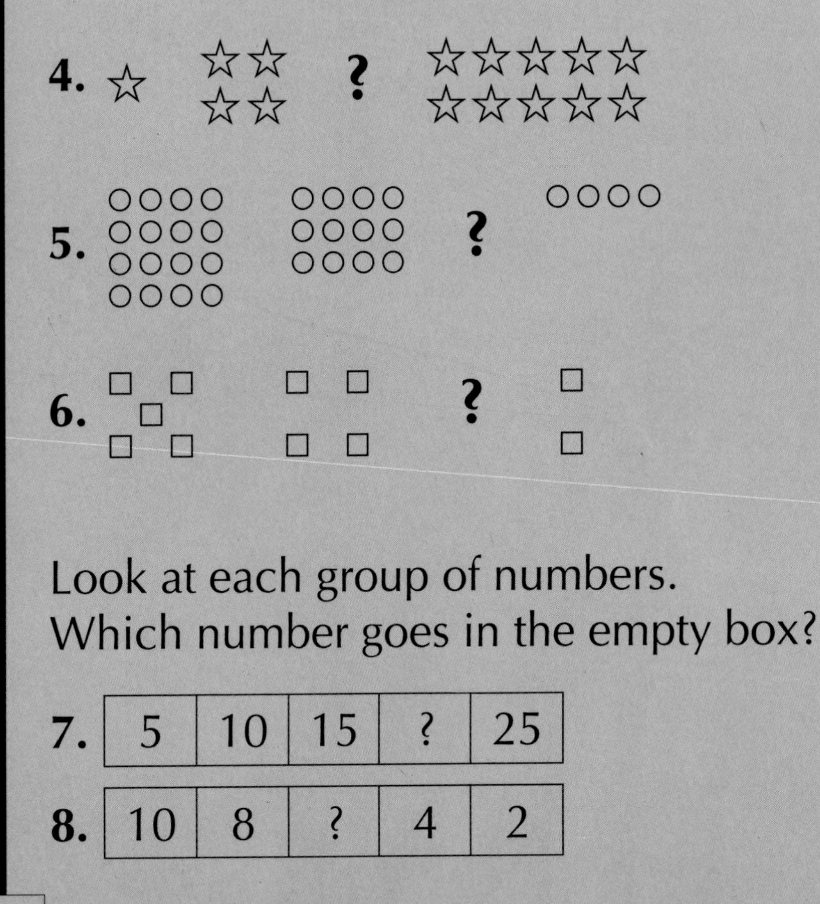

4. ☆ ☆☆☆☆ **?** ☆☆☆☆☆☆☆☆☆☆

5. **?**

6. **?**

Look at each group of numbers.
Which number goes in the empty box?

7.

5	10	15	?	25

8.

10	8	?	4	2

Answers
1. ▭
2. △
3. ◺
4. 7 stars
5. 8 circles
6. 3 squares
7. 20 **8.** 6

Look at each group of shapes.
Match the shapes to the puzzle they will make.

9.

a.

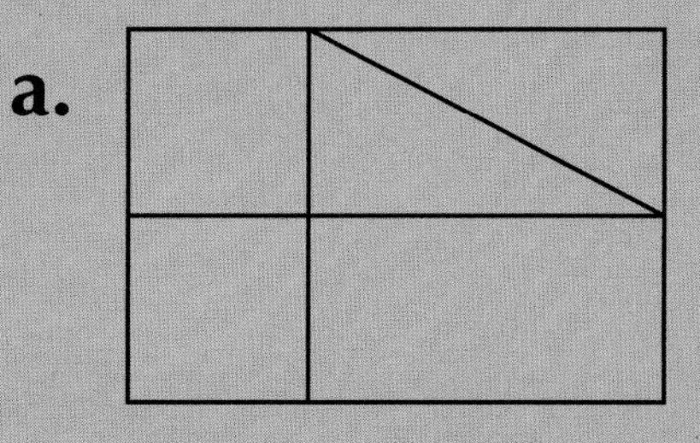

10.

b.

11.

c.

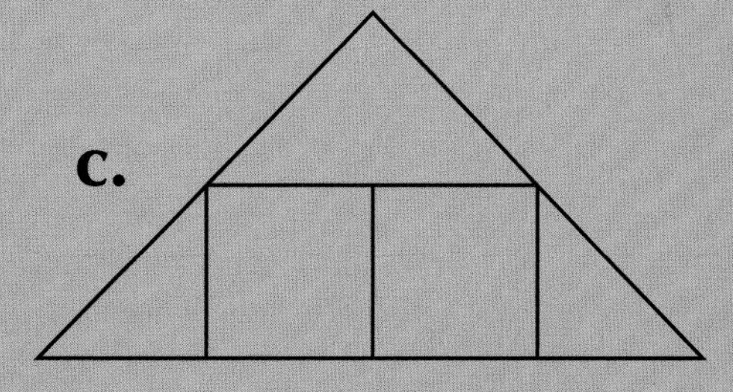

12.

d.

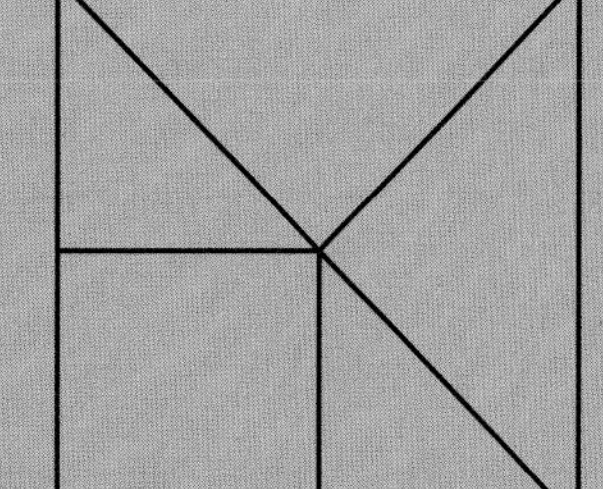

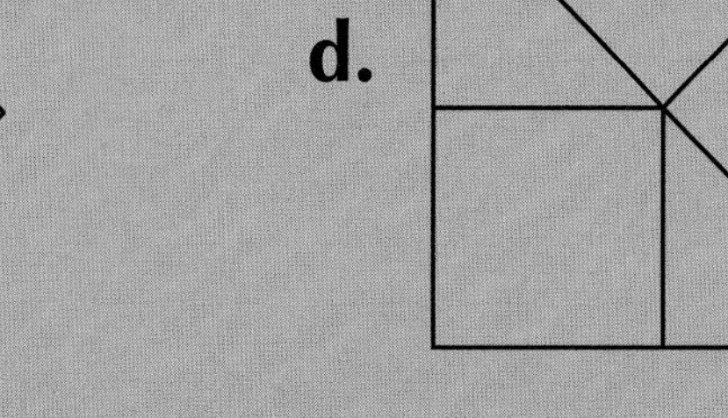

Answers **9.** b. **10.** d. **11.** a. **12.** c.